FLORA OF TROPICAL

ERIOSPERMACEAE*

Christopher Whitehouse

Herbs with perennial tubers and annual aerial parts, glabrous or partly pubescent (not in East Africa). Tubers globose to irregularly shaped; flesh white, yellow, pink or purplish red; growing points one or more, protected by a neck of sheathing fibrous remains of leaf-bases. Leaves 1–5, appearing with or after the flowers, erect to prostrate; blade linear to broadly ovate (sometimes circular or with enations in southern Africa), often with a distinct petiole-like base. Inflorescence a simple raceme, few–many-flowered, cylindrical to corymbose, dense to lax, upper flowers sometimes sterile or aborted; peduncle erect to suberect, sometimes subtended by a leaf-like sheath**; bracts minute, solitary, at the base of a single pedicel; pedicels short to long, erect to spreading, articulated at the apex. Flowers small, regular, rotate, campanulate or triangular, opening in the sun; tepals in 2 whorls of 3, connate at the base, subequal to equal or dimorphic, white, pink, yellow or green and often with a darker coloured midnerve, persistent. Stamens 6, hypogynous, joined to the base of the tepals; filaments filiform to broadly lanceolate, white, yellow, orange or purple blotched; anthers, dorsifixed, versatile, subglobose to oval, introrse. Ovary superior, sessile, with 3 locules, each with 2–6 axile ovules; style terete; stigma apical, small. Fruit a loculicidal capsule, turbinate to ovoid. Seeds oval to comma-shaped, densely covered with hairs, silvery white when fresh, turning brown with age.

The family comprises a single genus, with 102 species according to the recent revision, found only in Africa south of the Sahara, with a concentration in the western Cape Province of South Africa. It is easily recognised in fruit by the densely hairy seeds, and also by the bristly neck of old leaf-sheaths and leaves with a petiole-like base. It has little economic, medicinal or horticultural value but is botanically interesting for being a very isolated and advanced family, as discussed by Lu (1985).

ERIOSPERMUM

Willd., Sp. Pl., ed. 4, 2(1): 110 (1799); A.M. Lu in Nordic Journ. Bot. 5: 229–240 (1985); P.L. Perry in Contr. Bolus Herb. 17: 1–320 (1994)

Characters of the family.

Perry divided the genus into 3 subgenera and 9 sections based mainly upon the flower structure. Subgenus *Eriospermum* has distinctly dimorphic tepals and is restricted to South Africa and Namibia.

---

* This account is based largely upon P.L. Perry, A revision of the genus *Eriospermum* (Eriospermaceae), Contr. Bolus Herb. 17 (1994).
** Peduncle measurements are all to the first pedicel. The sheath is usually referred to as the peduncular bract.

subgen. **Ligulatum** *P.L. Perry* in Contr. Bolus Herb. 17: 19 (1994)

Flowers campanulate, becoming spreading to recurved in full sun. Tepals ligulate, equal or subequal, yellow or white, occasionally with a pink or blue tinge, often with a green to red midnerve.

sect. **Synanthum** *P.L. Perry* in Contr. Bolus Herb. 17: 32 (1994)

Plants normally flowering with the leaves. Tuber internally pink to red. Leaves usually 2 or more. *E. mackenii, E. triphyllum.*

sect. **Ligulatum** *P.L. Perry* in Contr. Bolus Herb. 17: 77 (1994)

Plants usually flowering before the leaves. Tuber internally white to pink. Leaves solitary. *E. abyssinicum.*

subgen. **Cyathiflorum** *P.L. Perry* in Contr. Bolus Herb. 17: 20 (1994)

Plants usually flowering before the leaves, though they may have a leaf-like peduncular bract. Tuber internally white. Flowers bowl-shaped to rotate. Tepals subequal, narrowly or more broadly obovate, white to pale green, sometimes suffused with red.

sect. **Rotatum** *P.L. Perry* in Contr. Bolus Herb. 17: 137 (1994)

Flowers flat, open or rotate. *E. kiboense.*

1. Flesh of tuber white; leaves solitary, usually appearing after the flowers, though sometimes with a leaf-like peduncular bract · 2
   Flesh of tuber pink to dark red; leaves usually more than 1, appearing with the flowers · 3
2. Peduncular bract visible above ground, up to 5.5 cm.; raceme cylindrical to narrowly conical; pedicels less than 1.3 cm.; flowers white to pale green · · · · · · · ·   4. *E. kiboense*
   Peduncular bract rarely visible above ground; raceme very lax; pedicels 1–18 cm.; flowers yellow · · · · · · ·   3. *E. abyssinicum*
3. Robust plant up to 50 cm. at flowering; leaves ovate to broadly ovate, usually less than $3\frac{1}{2}$ times as long as broad, coriaceous to fleshy · · · · · · · · · · · · · · · · · ·   1. *E. mackenii* (subsp. *mackenii*)

   Plant usually less than 27 cm. at flowering; leaves linear to lanceolate, usually more than $3\frac{1}{2}$ times as long as broad, slightly fleshy ·   2. *E. triphyllum*

1. **E. mackenii** (*Hook.f.*) *Baker* in J.L.S. 15: 266 (1876) & in Fl. Cap. 6: 378 (1896); P.L. Perry in Contr. Bolus Herb. 17: 33 (1994). Type: South Africa, Natal, cultivated in Hort. Kew., *Macken* (K, holo.!)

Plants 20–40(–50) cm., flowering with the leaves. Tuber up to 4 cm. long, 4.5 cm. wide, subglobose; skin dark brown, rough; flesh deep pink to maroon red. Old leaf-sheaths up to 6.5 cm. long. Leaves 2–5, erect to suberect; leaf-sheath barely exserted; blade narrowly to broadly ovate, 4–15 cm. long, 1.5–6.2 cm. wide, basally channelled and clasping the inner sheaths, bright green, coriaceous to fleshy, with raised veins. Peduncle up to 40 cm. long and 4 mm. diameter; raceme up to 24 cm. long, up to 3.5 cm. wide, cylindrical to narrowly conical, with up to 50 flowers; bracts ovate, attenuate, 1–2 mm. long, membranous; pedicels up to 2.5 cm. long, lower longer than upper. Flowers up to 2.1 cm. across, campanulate. Tepals ligulate, 5–10 mm. long, equal, bright yellow with a green midnerve, becoming spreading to recurved.

Filaments up to 6 mm. long, yellow, erect; anthers ± 0.5 mm., yellow. Ovary ovoid, pale green, 1.5–2.5 mm. long, 1–2 mm. wide; style cylindrical, 3–4.5 mm. long, white. Fruit obovoid, up to 8 mm. long and 7 mm. wide.

subsp. **mackenii**; P.L. Perry in Contr. Bolus Herb. 17: 34 (1994)

Plant entirely glabrous. Leaf-blade ovate to broadly ovate, usually less than 3¹/₂ times as long as broad.

TANZANIA. Kigoma/Tabora/Mpanda District: Ugalla R., cult. in Nairobi, *Bally* 8941!; Mpanda District: private road Mwesi [Mwese]–Mpanda, 53 km., 9 Dec. 1956, *Richards* 7191!; Chunya District: 1 km. W. of Itigi–Mbeya road, 29 Jan. 1969, *Edong* in *CAWM* 4469!
DISTR. **T** 4, 5, 7; Zambia, Mozambique, Zimbabwe and South Africa (Transvaal, Natal and eastern Cape Province)
HAB. Grassland in damp places; 1200–2000 m.

SYN. *Bulbine mackenii* Hook.f. in Bot. Mag. 28, t. 5955 (1872)
    E. junodii Baker in Bull. Herb. Boiss., sér. 2, 1: 783 (1901), *non* Baker (1904). Type: South Africa, Natal, Pinetown, *Junod* 149 (Z, holo.)
    E. veratriforme Poelln. in F.R. 52: 121 (1943). Type: Tanzania, Dodoma District, Uyansi, between Chaya [Tschaya] and Tura, Dec. 1926, *Peter* 34239 (B, holo.)

NOTE. Several of the East African specimens that key out to *E. mackenii* according to Perry's description, I believe to be no more than robust forms of *E. triphyllum*. In her revision, Perry stated that it was difficult to delimit the species in sect. *Synanthum* and therefore only gave a tentative arrangement until further study had been carried out on the tropical and subtropical species. On the face of it, I believe that her delimitation of *E. triphyllum* is too narrow, and some of the specimens that she cites (e.g. *Padwa* 70!*, *Richards* 15540!) do not fit her description. I have therefore taken a broader view of *E. triphyllum* and believe that the robustness of the plants is not a strong enough distinguishing feature, preferring the difference in leaf-shape, though this is still not infallible.
    Two collections from **T** 4 (*Bullock* 3260! & 1953!) appear to be this species but have 1, rarely 2, leaves up to 4 cm. long and the inflorescence is only 5–18 cm. long. They are not just young plants as the tubers are large and have a well-developed neck of leaf-sheaths.

2. **E. triphyllum** *Baker* in E.J. 15: 471 (1892); P.O.A. C: 140 (1895); Baker in F.T.A. 7: 473 (1898); Polhill in Journ. E. Afr. Nat. Hist. Soc. 24: 19, fig. 16 (1962); U.K.W.F., ed. 2: 313, t. 142 (1994); P.L. Perry in Contr. Bolus Herb. 17: 39 (1994). Type: Kenya, Machakos/Kitui District, Ukamba, *Hildebrandt* 2651 (K, holo.!, BM, iso.!)

Plants 4–27(–38) cm., flowering with the leaves. Tuber up to 3 cm. long, 3 cm. wide, globose to subglobose; skin pale fawn, rough; flesh pink becoming darker at the base. Old leaf-sheaths up to 5.5 cm. long. Leaves 1–3(–7), erect to suberect; sheath exserted up to 1.5 cm., green or mottled with maroon towards the base; blade linear to lanceolate or somewhat falcate, 1.5–15 cm. long, up to 2.5(–3.5) cm. wide, base cuneate and folded inwards, apex acute, bright green, soft, slightly fleshy. Peduncle 3.5–12(–20) cm. long, up to 2 mm. diameter; raceme 2–20.5 cm. long, up to 3 cm. wide, suberect, lax, with up to 50 flowers; bracts cymbiform, 1–1.5 mm. long, membranous, green becoming brown; pedicels up to 2.5 cm. long, lower longer than upper. Flowers 3–16 mm. across, campanulate. Tepals ligulate, 4–8 mm. long, subequal, yellow with red stippling on the midnerve, becoming spreading to recurved. Filaments up to 5 mm. long, yellow, erect. Ovary ovoid, pale green, 1–2 mm. long, 1–2 mm. wide; style cylindrical, up to 4 mm. long, white. Fruit 6–10 mm. long, 4–7 mm. wide. Seed ± 5 mm. long, 2 mm. wide, hairs up to 7 mm. long.

* Appendix 1 of Perry's revision gives the species number as 1c, *E. mackenii* subsp. *phippsii*, though the specimen is labelled by her as *E. triphyllum*. This is clearly a typographical error as *Hildebrandt* 2651, the type specimen of *E. triphyllum*, is also given as species number 1c.

UGANDA. Acholi District: Paimol, Apr. 1943, *Purseglove* 1533! & Chua, Paimol, *Eggeling* 1761!; Karamoja District: Mt. Debasien, 4 May 1948, *J. Adamson* 456!

KENYA. Northern Frontier Province: Moyale, 22 Apr. 1952, *Gillett* 12883!; near E. Elgon, Apr. 1951, *Tweedie* 906!; Kwale District: between Samburu and Mackinnon Road, near Taru, 3 Sept. 1953, *Drummond & Hemsley* 4147!

TANZANIA. Masai District: 20 km. Kabaya–Kijungu, 14 Jan. 1965, *Leippert* 5428!; Mbulu District: Tarangire National Park, 7 Nov. 1972, *Richards* 27978!; Iringa District: Ruaha National Park, near Msembe, 27 Dec. 1971, *I. Bjørnstad* 394!

DISTR. **U** 1; **K** 1–4, 6, 7; **T** 2, 7; Ethiopia, Zambia, Mozambique and Zimbabwe

HAB. Rocky soil in grassland or mixed scrub, often in dampish depressions; 250–2150 m.

SYN. *E. linearifolium* Baker in F.T.A. 7: 473 (1898). Type: Kenya, Teita District, Mbuyuni, *Scott Elliot* 6203 (K, holo.!, BM, iso.!)

    E. heterophyllum Cufod., Miss. Biol. Borana, Racc. Bot.: 314 (1939). Type: Ethiopia, Neghelli, *Cufodontis* 197 (location of holotype not known, photograph in Miss. Biol. Borana, Racc. Bot.: fig. 103!)

NOTE. In her revision, Perry stated that *E. linearifolium* was insufficiently known but noted that it was similar to *E. triphyllum*, differing significantly only in flower colour. In Baker's original descriptions, he gives the flower colour of both species to be white, only distinguishing the two on leaf width. As the leaf width of *E. linearifolium* falls within the variation shown by *E. triphyllum*, I consider the two to be synonymous. The other specimen Perry attributed to *E. linearifolium* (*Eggeling* 2813!) has much longer bracts (2–4 mm.) and is probably *Ornithogalum gracillimum*.

3. **E. abyssinicum** *Baker* in J.L.S. 15: 263 (1876) & in F.T.A. 7: 471 (1898); Bullock et al. in K.B. 1933: 98 (1933); Polhill in Journ. E. Afr. Nat. Hist. Soc. 24: 19 (1962); F.W.T.A., ed. 2, 3: 94, fig. 350 (1968); Moriarty, Wild Fl. Malawi: 29, t. 15/1 (1975); Vollesen in Opera Bot. 59: 89 (1980); Champl. & Maquet in Fl. Rwanda 4: 48 (1987); U.K.W.F., ed. 2: 313, t. 142 (1994); P.L. Perry in Contr. Bolus Herb. 17: 80 (1994). Lectotype, chosen by P.L. Perry (1994): Ethiopia, Gallabat, Gendua, *Schweinfurth* 26 (K, lecto.!, BM!, G, isolecto.)

Plants (4–)11–62 cm. (see note), usually flowering before the leaves. Tuber depressed-globose, up to 4.2 cm. long, 4 cm. wide; skin dark brown, thick, rough and flaky; flesh tough, corky, white. Old leaf-sheaths up to 6 cm. long. Leaf solitary, erect; sheath exserted up to 13 cm., 1–2 mm. diameter, terete, straight or slightly coiled, purple at the base, becoming glaucous towards the blade; blade narrowly lanceolate to falcate, cuneate, acuminate, mucronate, 5.5–14(–19) cm. long, up to 1.7 cm. wide, glaucous green, coriaceous, with prominent parallel veins, margins thickened or involute. Peduncle 2.5–18(–40) cm. long, up to 3 mm. diameter, dark red towards the base, green above; peduncular bract rarely observed above ground; raceme 7–32 cm. long, up to 10 cm. wide, very lax to dense (see note), conical to corymbose, with 10–50 flowers; bracts triangular, 1–2 mm. long, membranous with a brown midnerve; lower pedicels 1.5–18 cm. long, upper often much shorter. Flowers 3–14 mm. across, campanulate to spreading or recurved. Tepals 3–10 mm. long, subequal, lemon to bright yellow with a red-streaked green midnerve. Filaments subequal, filiform to narrowly lanceolate, yellow, inner erect, 3–4 mm. long, outer more spreading, 2–3 mm. long; anthers small, ovoid. Ovary ovoid, pale green with some red stippling, 2–2.5 mm. long, 1–2 mm. wide; style short, 1–2.5 mm. long, yellow. Fruit turbinate, 6–10 mm. long, 5–7 mm. wide. Seeds ± 6 mm. long, 2 mm. wide, hairs up to 7 mm. long.

UGANDA. W. Nile District: Arua, Mar. 1938, *Hazel* 454!; Acholi District: Naam, Apr. 1943, *Purseglove* 1519!; Bunyoro District: near Masindi, 16 Mar. 1907, *Bagshawe* 1543!

KENYA. Trans-Nzoia District: Moi's [Hoey's] Bridge, Mar. 1936, *Mainwaring* 20!; Uasin Gishu District: near Kapsaret [Kaposoret] Forest Reserve, 8 May 1951, *G.R. Williams* 171!; S. Elgon, Mar. 1937, *Tweedie* 380!

TANZANIA. Musoma District: 91 km. Seronera to Klein's Camp, 6 Apr. 1961, *Greenway & Myles Turner* 9994!; Iringa District: Ruaha National Park, 1 km. E. of Msembe, 10 Dec. 1971, *A.*

FIG. 1. *ERIOSPERMUM KIBOENSE* – **1**, plant with tuber and leaf, × 1; **2**, leaf, face view, × 1; **3**, peduncle and peduncular bract, × 1; **4**, inflorescence, × 1; **5**, flower, face view, × 4.5; **6**, flower, side view, × 4.5; **7**, outer tepal and stamen, × 4.5; **8**, inner tepal and stamen, × 4.5; **9**, gynoecium, × 4.5. Reproduced from Contr. Bolus Herb. 17: fig. 75 (1994).

Bjørnstad 1138!; Songea District: 21 km. N. of Songea, by Lumecha Bridge, 3 Jan. 1956, *Milne-Redhead & Taylor* 8035!

DISTR. **U** 1, 2; **K** 3, 5; **T** 1, 2, 4, 6–8; widespread in tropical and subtropical Africa

HAB. Rocky soil in grassland or open woodland, often in damp areas; 150–2450(–2750) m.

SYN. *Bulbine unifolia* Schweinf., Pl. Gallab. Exsic. No. 26 (1865), *nomen nudum*, name only on herbarium specimen

  Eriospermum fleckii Schinz in Bull. Herb. Boiss., sér. 1, 4, App. 3: 37 (1896). Type: Namibia, Rehoboth, *Fleck* 887 (Z, holo.)

  E. luteorubrum Baker in Fl. Cap. 6: 372 (1896). Lectotype, chosen here: South Africa, Transvaal, Barberton, summit of Saddleback Range, *Galpin* 528 (K, lecto.!, BOL, PRE, SAM, Z, isolecto.)

  E. burchellii Baker in Fl. Cap. 6: 372 (1896). Lectotype, chosen by P.L. Perry (1994): South Africa, Griqualand West, near the Asbestos Mountains, between Wittewater and Rietfontein, *Burchell* 2008 (K, lecto.!)

  E. elatum Baker in F.T.A. 7: 471 (1898). Lectotype, chosen by P.L. Perry (1994): Zambia, Urungu, *Carson* 16 (K, lecto.!)

  E. schinzii Engl. & K. Krause in E.J. 45: 140 (1910), *nom. illegit.*, *non* Baker (1904). Type: Namibia, Grootfontein, *Dinter* 923 (SAM, iso.)

NOTE. *E. abyssinicum* is the most widespread species and highly variable. As a result there are many local variants, which upon further study may deserve taxonomic ranking. A dwarf form is found in **T** 4 and **T** 7, on burnt grassland, 1650–2750 m. (*Richards* 13031!, *Richards* 18454!, *Procter* 1417!). It only grows 4–8 cm. high and has a dense inflorescence with 18–50 flowers. Intermediate specimens have been found in Malawi, growing 6.5–13 cm. high, but they have a laxer raceme.

4. **E. kiboense** *K. Krause* in E.J. 48: 356 (1912); P.L. Perry in Contr. Bolus Herb. 17: 140 (1994). Type: Tanzania, Kilimanjaro, *Endlich* 711 (M, holo.)

Plants 16–55 cm., flowering before the leaves. Tuber pyriform to irregular, up to 5 cm. long, 3.5 cm. wide; skin fawn, tough and smooth; flesh white. Old leaf-sheaths up to 9 cm. long. Leaf solitary, suberect; sheath exserted up to 6 cm., ± 3 mm. diameter, terete, reddish purple; blade broadly elliptic to lanceolate, up to 14 cm. long, 9 cm. wide, bright green above, paler beneath, coriaceous, margins entire. Peduncle 14–45 cm. long, 2–3 mm. diameter, red tinged towards the base, glaucous green above; peduncular bract exserted 4–7 cm., lower part sheathing, becoming leaf-like, broadly ovate, up to 4 cm. long, 3 cm. wide, acute; raceme 3–16 cm. long, up to 3 cm. wide, cylindrical to narrowly conical, with 10–50 flowers; bracts subovate, acuminate, somewhat saccate, ± 1 mm. long, membranous, colourless; pedicels up to 1.3 cm. long. Flowers 3–10 mm. across, rotate. Tepals 3–6 mm. long, 2 mm. wide, equal to subequal, white to very pale green with a darker green to brown midnerve. Filaments subequal, oblong, erect, 2.5–3 mm. long; anthers small, green. Ovary ovoid to globose, bright green, 1–2 mm. long, 1–2 mm. wide; style cylindrical, ± 1.5 mm. long, white. Fruit up to 9 mm. long, 9 mm. wide. Seeds 2–3 mm. long, hairs up to 7 mm. long. Fig. 1.

TANZANIA. Arusha District: Songe Hill, by the telegraph station, 23 Feb. 1969, *Richards* 24175!; Moshi District: Engare [Ngare] Nairobi, Dec. 1965, *Beesley* 189!; Ufipa District: Sumbawanga, Mbisi [Mbesi] Forest, 11 Nov. 1963, *Richards* 18400!

DISTR. **T** 2, 4, 7; Malawi and Zimbabwe

HAB. Grassland or semi-shade in open woodland; 1000–2250 m.

SYN. *E. erectum* Suess. in Trans. Rhod. Sci. Assoc. 43: 74 (1951). Type: Zimbabwe, Marandellas, *Dehn* 499 (M, holo.)

INDEX TO ERIOSPERMACEAE

Printed and bound by CPI Group (UK) Ltd, Croydon, CR0 4YY

23/10/2024

01777676-0002